The Organic Chemistry Cookbook

An Original Organic Chemistry Study Aid

by Chef Fred

The Organic Chemistry Cookbook
Copyright © 2000

Table of Contents

Introduction

Introduction

The Organic Chemistry Cookbook contains reactions commonly encountered in an introductory organic chemistry course. It is best used as an accompaniment to an organic chemistry text, for a working knowledge of nomenclature and fundamental reaction mechanics is assumed. Unique in design, *The Organic Chemistry Cookbook* presents core concepts within the framework of lab synthesis of organic compounds.

Each reaction presented in *The Organic Chemistry Cookbook* gives an example of how to produce a specific organic compound, or *dish*. In typical cookbook format, each dish is presented with a list of ingredients (reactants) and method of preparation. The *Ingredients* and *Synthesis Synopsis* sections represent a general* summary for making the desired product whether it be an alkane, alcohol, or ß-ketoester, etc.

The sections *Reaction Mechanism* and *Explanation of Steps* explore the results which follow *Ingredients* and *Synthesis Synopsis*. Unlike the typical illustration of a reaction mechanism presented in a textbook, double-sided arrows indicating a reversible step are not used; yet, reversible steps that merit examination may be addressed. Moreover, illustrations of a mechanism are hand drawn for a more *organic* appeal.

All dishes close with a question and answer section entitled *Ask the Chef*. *Ask the Chef* provides further analysis of the reaction mechanism, addressing questions typically concerned with key steps of the mechanism as well as the consequences of variations to the list of ingredients.

---

* Ingredients and Synthesis Synopsis serve only to highlight the cookbook theme. Specific amounts of starting reactants as well as a detailed and accurate depiction of the synthesis are beyond the scope of this book.

Dishes

Alcohol
by an S_N1 Reaction

2-Methyl-2-Butanol

2-Bromo-3-Methylbutane → 2-Methyl-2-Butanol

Ingredients:

Water (polar protic solvent and nucleophile)
2-Bromo-3-Methylbutane (alkyl halide)

Synthesis Synopsis:

Add 2-bromo-2-methylbutane and water to an Erlenmeyer flask. Heat flask over hot plate. Distill to purify.

Reaction Mechanism:
S_N1 Reaction

Step 1	
Step 2	
Step 3	
Step 4	

Explanation of Steps:

(1) A secondary carbocation and a bromide ion are formed as the halogen (retaining a pair of electrons) is cleaved heterolytically from the alkyl halide.

(2) Rearrangement occurs as a hydride shift furnishes a more stable tertiary carbocation.

(3) An alkoxonium ion (the conjugate acid of an alcohol) is formed as nucleophilic water attacks the carbocation. In the process, oxygen gains a positive charge.

(4) The alkoxonium ion is easily deprotonated by water (acting as a Bronsted base).

Ask the Chef:

(1) *Would the use of a nonpolar solvent, e.g., toluene, in place of water produce the same results?*

Chef: No. The energy needed to produce two charged species from heterolytic cleavage of the carbon-halogen bond (Step 1) stems from waters ability to separate the positive and negative charges of the resulting ions (a carbocation and bromide ion) by solvation. In general, polar solvents, such as water, serve to increase the rate of an S_N1 reaction, whereas nonpolar solvents (those with poor solvation ability), such as toluene, are of no use in an S_N1 reaction.

(2) *What effect, if any, would substituting water with another polar protic solvent, e.g., methanol, have on the course of the reaction?*

Chef: Substituting methanol for water would lead to an ether, instead of an alcohol product. Methanol would serve to solvate the ions in solution and attack the resulting carbocation. Subsequent deprotonation by methanol would result in formation of the ether product.

Alcohol
by an S_N2 Reaction

2-Butanol

2-Bromobutane   2-Butanol

Ingredients:

2-Bromobutane (alkyl halide)
Saturated sodium hydroxide (nucleophilic base)

Synthesis Synopsis:

Add 2-bromobutane and saturated sodium hydroxide to an Erlenmeyer flask. Heat flask over hot plate. Distill to purify.

Reaction Mechanism: S_N2 **Reaction**		
Step 1	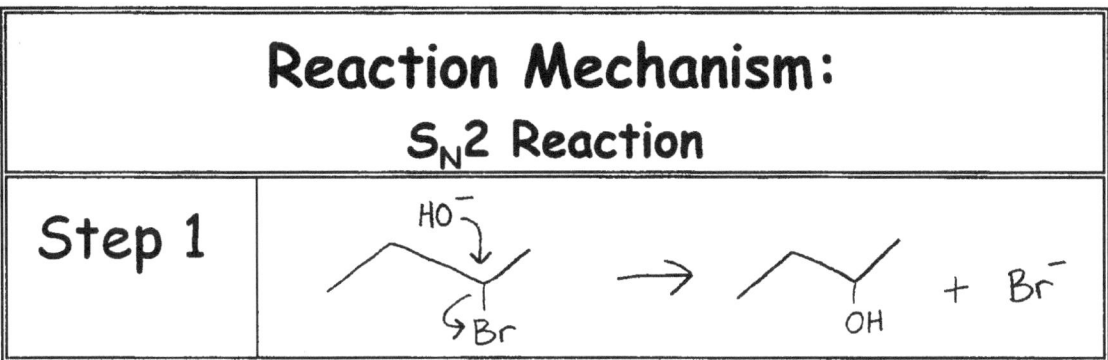

Explanation of Steps:

(1) In a concerted S_N2 fashion, nucleophilic hydroxide ion attacks the carbon atom bearing the leaving group, producing the alcohol product and halide ion.

Ask the Chef:

(1) *What effect, if any, would substituting bromine (the leaving group) with another halogen have on the course of the reaction?*

<u>Chef</u>: Changing the leaving group would affect the rate of the reaction by either slowing, or speeding up the rate-determining ionization step. A halogen's ability to be displaced correlates to the acidity of its conjugate acid. Therefore, I⁻ (the weakest base of the halogens) is the best leaving group of the halogens because HI is the strongest conjugate acid (has the lowest pKa value) of the group. Accordingly, F⁻ (the strongest base of the halogens) is quite difficult to displace in an S_N2 process.

Alkene
by an E1 Reaction

2-Methylpropene

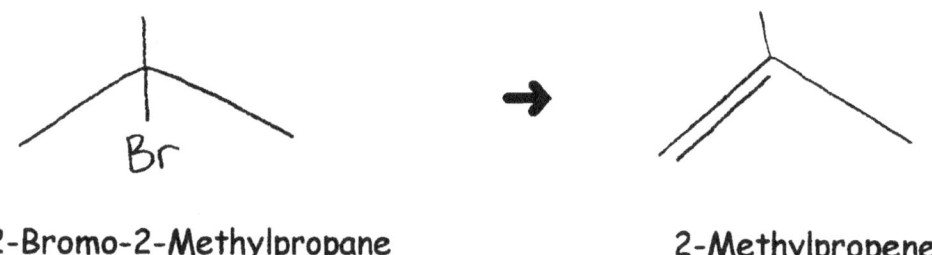

2-Bromo-2-Methylpropane   2-Methylpropene

Ingredients:

2-Bromo-2-Methylpropane (alkyl halide)
Methanol (polar protic solvent)

Synthesis Synopsis:

Add 2-bromo-2-methylpropane and methanol to an Erlenmeyer flask. Heat flask over hot plate. Extract and dry the organic product and distill to purify. Assess purity and yield of product by gas chromatography.

Reaction Mechanism: E1 Reaction	

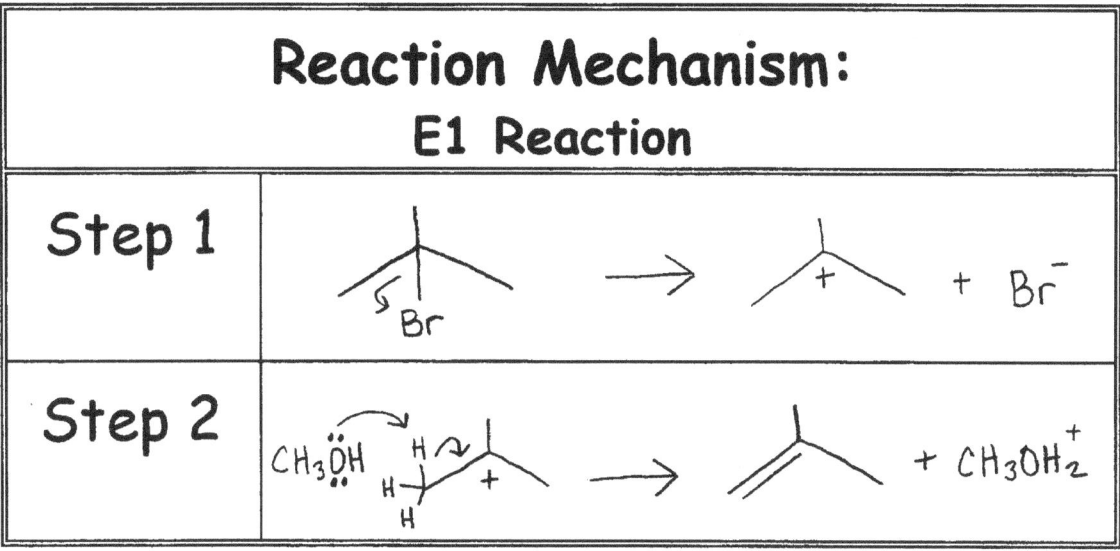

Step 1	
Step 2	

Explanation of Steps:

(1) A tertiary carbocation intermediate is formed upon ionization of the carbon-halogen bond.

(2) An alkene and an alkoxonium ion are produced as methanol deprotonates the carbocation and a double bond forms.

Ask the Chef:

(1) *Would substituting methanol with a nonpolar solvent, such as toluene, produce the same results?*

<u>Chef</u>: No. Analogous to an S_N1 process, the rate-determining step of the E1 reaction involves dissociation of the alkyl halide via solvolysis. A nonpolar solvent, such as toluene, lacks such solvation properties, and is therefore, of no use in E1 reactions.

(2) *Referring to Step 2, why doesn't methanol attack the positively charged carbon of the carbocation intermediate as seen in the S_N1 reaction?*

<u>Chef</u>: Following ionization of the carbon-halogen bond, methanol may act as a Bronsted base (by abstracting a proton from the carbocation intermediate), as seen in an E1 reaction, or as a nucleophile (by attack on the positively charged carbon), as seen in an S_N1 reaction. Competition between E1 and S_N1 reaction pathways are quite common, especially among secondary alkyl halides involving a good leaving group (such as -Br) and a moderately weak nucleophile/base. Therefore, a significant yield of S_N1 *ether* product (2-methoxy-2-methylpropane) is observed.

Alkene
by an E2 Reaction

2-Methyl-1-Butene

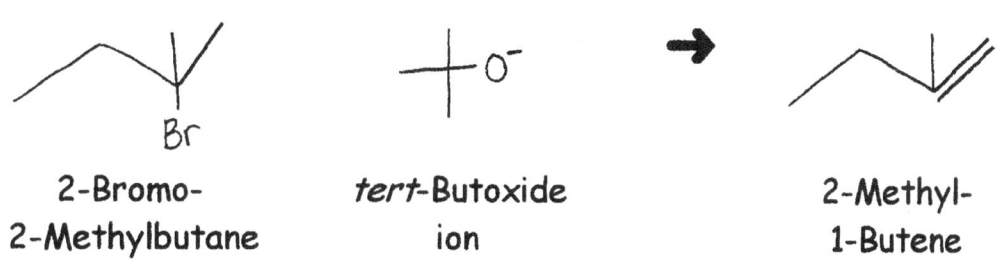

| 2-Bromo-
2-Methylbutane | *tert*-Butoxide
ion | 2-Methyl-
1-Butene |

Ingredients:

2-Bromo-2-Methylbutane (alkyl halide)
Potassium *tert*-Butoxide (Bronsted base)
tert-Butyl Alcohol (polar protic solvent)
(*tert*-butyl alcohol is the common name for 2-methyl-2-propanol)

Synthesis Synopsis:

Add 2-bromo-2-methylbutane, potassium *tert*-butoxide, and *tert*-butyl alcohol to an Erlenmeyer flask. Heat flask over hot plate. Extract and dry the organic product and distill to purify. Assess purity and yield of product by gas chromatography.

Reaction Mechanism: **E2 Reaction**	
Step 1	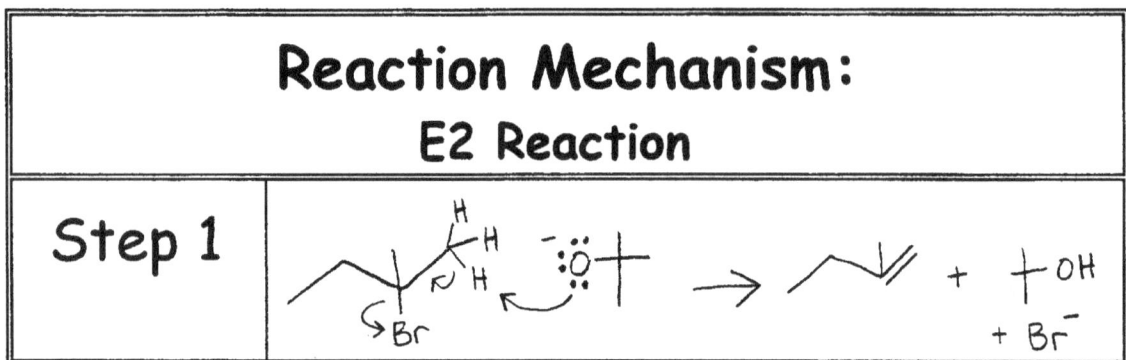

Explanation of Steps:

(1) In a concerted process, *tert*-butoxide ion deprotonates the alkyl halide as a double bond forms and the halogen departs.

Ask the Chef:

(1) *Would substituting potassium tert-butoxide with a less sterically hindered base, such as sodium hydroxide, produce the same results?*

<u>Chef</u>: The steric bulk of potassium *tert*-butoxide leads to deprotonation at the less hindered region of the alkyl halide, leading to a greater yield of 2-methyl-1-butene than 2-methyl-2-butene (the more highly substituted alkene). In the case of sodium hydroxide, the products are the same but the product ratio is *reversed* (2-methyl-2-butene predominates over 2-methyl-1-butene) Therefore, the regioselectivity of the base is primarily governed by it's steric bulk.

(2) *Is any S$_N$2 product observed?*

<u>Chef</u>: No. The steric bulk of the base as well as the steric hindrance of the substrate retards nucleophilic attack at the carbon bearing the leaving group.

*For comparison and contrast see the Dehydrohalogenation reation presented on page 18.

Alkene
by Alcohol Dehydration

Cyclohexene

Cyclohexanol   Cyclohexene

Ingredients:

Cyclohexanol (alcohol)
Concentrated sulfuric acid (acid catalyst)

Synthesis Synopsis:

Add cyclohexanol and sulfuric acid to a round-bottomed flask. Reflux. Extract and dry the organic product and distill to purify. Assess purity and yield of product by gas chromatography.

Reaction Mechanism:
Alcohol Dehydration

Step 1	
Step 2	
Step 3	

Explanation of Steps:

(1) Protonation weakens the C-O bond and converts the hydroxyl group into a good leaving group ($-OH_2^+$).

(2) A secondary carbocation is produced upon heterolytic cleavage of $-OH_2^+$.

(3) Water, acting as a Bronsted base, abstracts a proton from carbon adjacent to the positive center to furnish the alkene.

Ask the Chef:

(1) *Referring back to step 3, why is the positive center not attacked by HSO_4^- (the nucleophilic species in the surrounding media)?*

<u>Chef</u>: Hydrogen sulfate (HSO_4^-) is a weak base (poor nucleophile), therefore, the loss of an adjacent proton to water is the course by which the reaction proceeds.

(2) *Why is heat required?*

<u>Chef</u>: The dehydration reaction is slow. Heat favors the elimination of the hydrogen adjacent to the positive center to produce the alkene. Lowering the heat, or the absence of heat, will decrease the yield of the alkene product. In general, primary alcohols are more difficult to dehydrate than secondary alcohols which are more difficult to dehydrate than tertiary alcohols. Tertiary alcohols, which usually have the most stable transition state, dehydrate with relative ease.

(3) *Will dilute sulfuric acid produce the same results?*

<u>Chef</u>: No. Water is a byproduct of the reaction. Therefore, excess water will shift the equilibrium of the reaction toward the reactants, lowering the yield of product.

Alcohol
by Acid Catalyzed Hydration

2,3-Dimethyl-2-Butanol

3,3-Dimethyl-1-Butene   2,3-Dimethyl-2-Butanol

Ingredients:

3,3-Dimethyl-1-Butene (alkene)
Dilute sulfuric acid (acid catalyst)

Synthesis Synopsis:

Add the alkene and dilute acid to an Erlenmeyer flask. Swirl while
heating flask gently over hot plate. Distill to purify.

Reaction Mechanism:
Acid Catalyzed Hydration

Step 1	
Step 2	
Step 3	
Step 4	

Explanation of Steps:

(1) Water and a secondary carbocation are formed as pi electrons transfer a proton from the hydronium ion to the less substituted carbon of the double bond.

(2) Migration of a neighboring alkyl group to the positively charged carbon furnishes a more stable tertiary carbocation.

(3) Water, acting as a nucleophile, forms a bond with the positively charged carbon, producing an alkoxonium ion.

(4) The alcohol product is formed and a hydronium ion is regenerated as the alkoxonium ion is deprotonated by water.

Ask the Chef:

(1) *Referring to Step 1, why is the less highly substituted carbon of the double bond protonated?*

<u>Chef</u>: Protonation at the less highly substituted end of the double bond produces a more stable (secondary) carbocation than would result from protonation at the more highly substituted end of the double bond (a primary carbocation). Such an addition is said to be in accord with Markovnikov's rule.

(2) *What prompts the alkyl shift?*

<u>Chef</u>: The addition of a hydrogen ion to the less substituted carbon leads to a secondary carbocation intermediate. Migration of a methyl group with its bonding electrons from an adjacent carbon gives a more stable tertiary carbocation.

(3) *Can the positive center be attacked by water, before the alkyl shift occurs, to yield 3,3-dimethyl-2-butanol?*

<u>Chef</u>: Yes. In fact, 3,3-dimethyl-2-butanol is the minor product of the reaction. Rearrangement to a more stable carbocation, however, is greatly favored over straightforward hydration.

Alkene
by Dehydrohalogenation

2-Methyl-2-Butene

Sodium ethoxide   2-Bromo-2-Methylbutane   2-Methyl-2-Butene

Ingredients:

Ethanol (solvent)
2-Bromo-2-Methylbutane (alkyl halide)
Sodium ethoxide (base)

Synthesis Synopsis:

Add ethanol, 2-bromo-2-methylbutane, and sodium ethoxide to a round-bottomed flask. Reflux. Extract and dry the organic product. Distill to purify. Assess purity and yield of product by gas chromatography.

Reaction Mechanism: Dehydrohalogenation	
Step 1	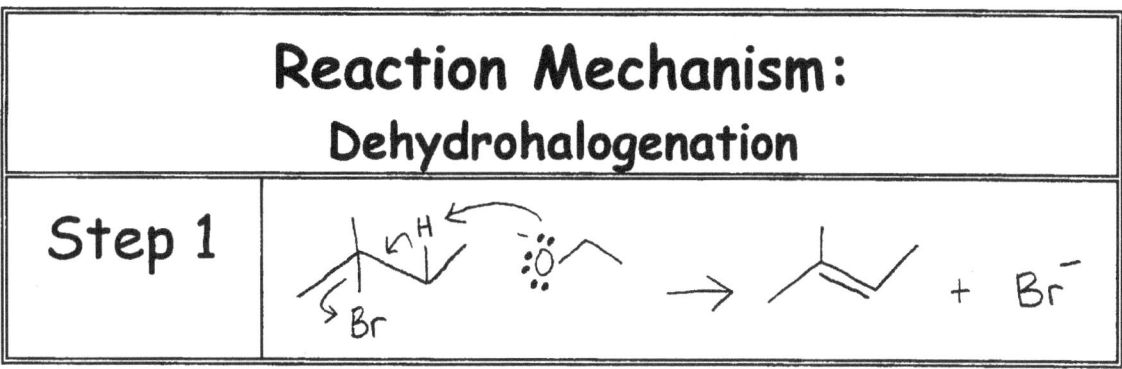

Explanation of Steps:

(1) In a typical concerted E2 reaction, a lone pair of electrons on oxygen of sodium ethoxide abstracts a proton from the carbon adjacent to the carbon atom bearing the leaving group (-Br). In the process, the leaving group departs and a double bond is formed.

Ask the Chef:

(1) *Does the E2 reaction route compete with an S_N2 pathway?*

<u>Chef</u>: The combination of a strongly basic unhindered nucleophile and Sterically hindered starting haloalkane (the methyl group attached to the carbon bearing the leaving group impedes nucleophilic attack) favors the E2 reaction route. Nonetheless, the use of such a strongly basic unhindered nucleophile leads to some S_N2 ether product.

(2) *Why is the more highly substituted alkene the favored product?*

<u>Chef</u>: In general, the more substituted the sp^2 *carbons* (the carbons forming the double bond), the more stable the alkene. Hence, 2-methyl-2-butene, containing three alkyl substituents, is favored over 2-methyl-1-butene (the minor product of the reaction), containing two alkyl substituents.

(3) *Would the use of potassium tert-butoxide in place of sodium ethoxide produce the same results?*

<u>Chef</u>: Sterically bulky bases, e.g., potassium *tert*-butoxide, in an E2 reaction, tend to deprotonate the less highly substituted end of the double bond, which leads to the less substituted alkene. Therefore, the less highly substituted alkene becomes the major product of the reaction.

*For comparison and contrast see the E2 reaction presented on page 9.

Ester
by Fischer Esterification

Pentyl Acetate

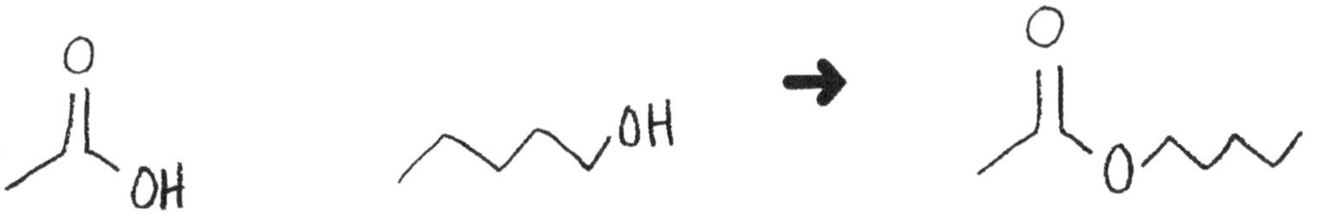

Acetic acid   1-Pentanol   Pentyl acetate

Ingredients:

Acetic acid (carboxylic acid)
1-Pentanol (alcohol)
Concentrated sulfuric acid (catalyst)

Synthesis Synopsis:

Add acetic acid, 1-pentanol, and concentrated sulfuric acid to a round-bottomed flask. Reflux. Extract and dry the organic product and distill to purify. Assess purity and yield of product by gas chromatography.

Reaction Mechanism:
Fischer Esterification

Step 1	
Resonance Structures:	
Step 2	
Step 3	
Step 4	
Step 5	
Resonance Structures:	

Reaction Mechanism: Fischer Esterification	
Step 6	

Explanation of Steps:

(1) A lone pair of electrons of the carbonyl oxygen of acetic acid abstracts a proton of the acid catalyst, putting the carbonyl oxygen in a positively charged state. Resonance forms are produced as valence electrons are redistributed to offset the charge on oxygen.

(2) Acting as a nucleophile, 1-pentanol attacks carbonyl carbon of the major resonance contributor. In the process, pi electrons migrate to the carbonyl oxygen.

(3) A neutral tetrahedral intermediate is produced as positively charged oxygen breaks its bond with hydrogen.

(4) A hydroxyl group of the tetrahedral intermediate is protonated, turning a bad leaving group (-OH) into a good leaving group ($-OH_2^+$).

(5) The $-OH_2^+$ group is displaced (water is generated) as a lone pair of electrons of hydroxyl oxygen forms a double bond with carbon. Resonance forms are produced as valence electrons are redistributed to offset the charge on oxygen.

(6) Positively charged oxygen breaks its bond with hydrogen and gains a pair of electrons, furnishing the ester.

Ask the Chef:

(1) *Referring back to Step 1, why does the carbonyl oxygen grab the hydrogen ion in solution instead of the hydroxyl oxygen?*

<u>Chef</u>: The addition of hydrogen on the carbonyl group gives a resonance stabilized intermediate.

(2) *Will a dilute acid catalyst produce the same results?*

<u>Chef</u>: No. Refluxing the ester with dilute acid means refluxing the ester with more water (a byproduct of the reaction). Excess water will shift the equilibrium of the reaction towards the reactants, lowering the yield of the ester product.

(3) *Referring to Step 4, why does protonation take place at the -OH group (to furnish water) rather than the -OR group (to furnish pentanol)?*

<u>Chef</u>: The reaction route which leads to product is that which involves the protonation at the -OH group. Nonetheless, protonation at the -OR group occurs more readily than protonation at the -OH group since pentanol is a weaker base, and therefore, better leaving group than water. Therefore, an excess amount of carboxylic acid is used to improve the yield of the ester.

Alcohol
by Ester Reduction

2-Phenyl-1-Propanol

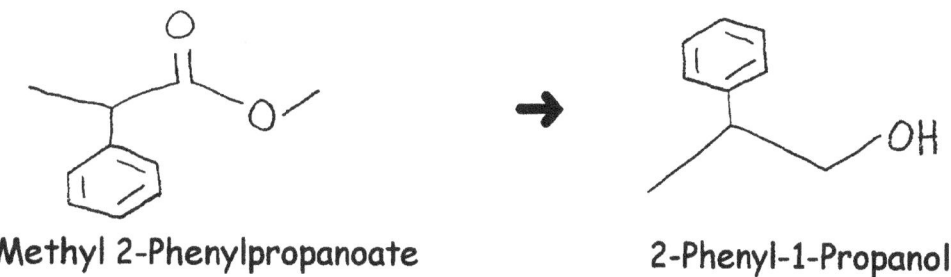

Methyl 2-Phenylpropanoate   2-Phenyl-1-Propanol

Ingredients:

Ether (solvent)
Methyl 2-Phenylpropanoate (ester)
Lithium Aluminum Hydride (reducing agent)
Dilute hydrochloric acid (for aqueous work-up)

Synthesis Synopsis:

Under nitrogen add methyl 2-phenylpropanoate, lithium aluminum
hydride, and ether to an Erlenmeyer flask. Incubate flask in hot water
bath. Let solution cool. Add dilute hydrochloric acid. Distill to purify.

Reaction Mechanism:
Ester Reduction

Step 1	
Step 2	
Step 3	
Step 4	

Explanation of Steps:

(1) In a concerted fashion, a hydride ion (supplied by lithium aluminum hydride) is transferred to the partially positive carbonyl carbon as pi electrons migrate to the carbonyl oxygen, producing a tetrahedral carbonyl addition intermediate.

(2) The tetrahedral carbonyl addition intermediate collapses as $-OCH_3$ is lost and a carbonyl bond is regenerated, furnishing an aldehyde.

(3) An alkoxide ion is formed as lithium aluminum hydride transfers a second hydride ion to the carbonyl carbon and pi electrons migrate to oxygen.

(4) Upon aqueous acid work-up, negatively charged oxygen is protonated and the alcohol product is formed.

Ask the Chef:

(1) *Why does the reaction proceed from the aldehyde product of Step 2?*

Chef: Two major factors contribute to the progressive aldehyde production step: (1) Lithium aluminum hydride is a strong reducing agent, and (2) aldehydes are more reactive toward nucleophilic attack than esters.

(2) *Can the reaction be stopped at the aldehyde production step?*

Chef: A mild reducing agent must be used to stop the reduction at the aldehyde stage. Diisobutylaluminum hydride (DIBAL, or DIBAH) is a mild reducing agent commonly employed to convert esters into aldehydes. The reducing power of DIBAL is primarily governed by its steric bulk, which hinders easy transfer of hydride ions.

Alkene
by the Wittig Reaction

Stilbene

| Ylide | Benzaldehyde | E-Stilbene | Z-Stilbene |

Ingredients:

Benzaldehyde (aldehyde)
Triphenylphosphine
Benzyl chloride (alkyl halide)
Butyllithium (Bronsted base)

Synthesis Synopsis:

Under nitrogen add triphenylphosphine and benzyl chloride to an
Erlenmeyer flask. Heat flask gently over hot plate. Add butyllithium
and stir. Add benzaldehyde. Extract and dry the organic products.

Reaction Mechanism:
Wittig Reaction

Step 1	
Step 2	
Resonance Structures:	
Step 3	
Step 4	
Step 5	

Explanation of Steps:

(1) In an S_N2 fashion, triphenylphosphine attacks benzyl chloride, producing triphenylbenzylphosphonium chloride, a phosphonium salt. The positive charge on phosphorus renders the hydrogens of the adjacent carbon acidic and ready targets for nucleophilic attack.

(2) Resonance structures of the phosphorus ylide are produced as the nucleophilic carbon of butyllithium abstracts a proton from the carbon adjacent to the positively charged phosphorus.

(3) A phosphorus betaine (a zwitterion) is formed as negatively charged carbon of the ylide attacks the carbonyl carbon of benzaldehyde.

(4) Bond formation between negatively charged oxygen and positively charged phosphorus forms a four-membered oxaphosphatane ring.

(5) Collapse of the four-membered ring produces the alkene product and triphenylphosphine oxide.

Ask the Chef:

(1) *Are the isomers of the stilbene product produced in equal amount?*

<u>Chef</u>: No. Analysis of product shows that the E form predominates. The amount of E-stilbene formed is greater than Z-stilbene for steric reasons.

(2) *Referring to Step 1, it is shown that triphenylphosphine attacks a primary alkyl halide in an S_N2 fashion. What effect, if any, does the use of secondary or tertiary alkyl halides have on the course of the reaction?*

<u>Chef</u>: Like any S_N2 reaction, the success of nucleophile attack depends on the bulk of the target molecule. Therefore, triphenylphosphine reacts best with methyl and primary alkyl halides. Attack on secondary alkyl halides is slower and product yield is reduced. Tertiary alkyl halides, due to their resistence to nucleophilic attack are least desired in the wittig reaction. Accordingly, the ylide is the usual carrier of the less bulky substituent of the double bond.

(3) *Is the Wittig reaction used to convert the carbonyl bond of esters into a carbon-carbon double bond to produce the corresponding enol product?*

<u>Chef</u>: Although the Wittig reaction is a valuable means of converting the carbonyl bonds of aldehydes and ketones to their corresponding alkene product, it is not employed in the conversion of esters to enols. The difference of polarity of the carbonyl carbon of esters makes it not as ready a center for ylide attack as the carbonyl carbon of aldehydes and ketones.

(4) *What advantage, if any, does the Wittig reaction have over the E2 dehydrohalogenation of alkyl halides in the synthesis of alkenes?*

<u>Chef</u>: The difference between the two routes of synthesis lies in the regioselectivity of each reaction. The Wittig reaction is completely regioselective. Therefore, the location of the double bond of the alkene product can be predicted with certainty. E2 elimination reactions typically produce more than one alkene product—the major product being the most substituted alkene in most cases (especially when the use of a bulky base is avoided).

Alcohol
by the Grignard Reaction

Triphenylmethanol

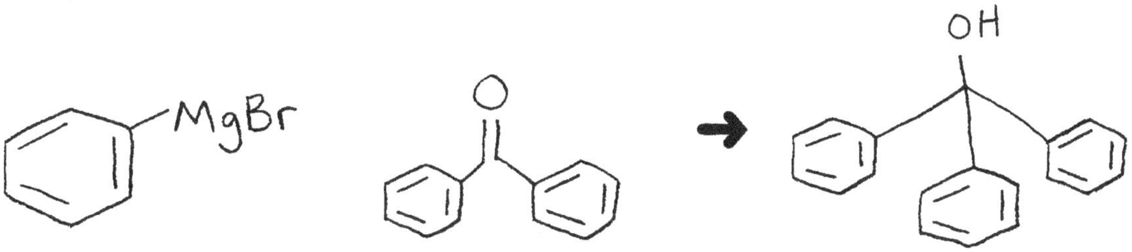

| Phenylmagnesium bromide | Benzophenone | Triphenylmethanol |

Ingredients:

Phenylmagnesium bromide (Grignard reagent)
Benzophenone (ketone)
Anhydrous diethyl ether (solvent)
Dilute sulfuric acid (for aqueous work-up)

Synthesis Synopsis:

Add benzophenone, phenylmagnesium bromide, and anhydrous diethyl ether to a round-bottomed flask. Reflux. Add dilute sulfuric acid. Extract and dry the organic product. Record mass and melting point of precipitated product.

Reaction Mechanism: Grignard Reaction	
Step 1	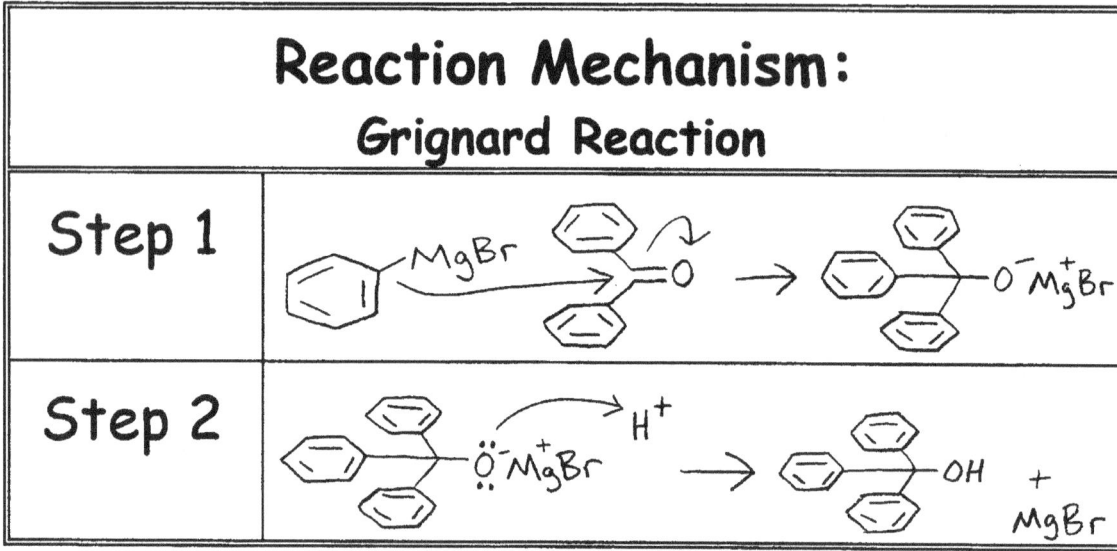
Step 2	

Explanation of Steps:

(1) Nucleophilic carbon of the Grignard reagent's polar covalent carbon-magnesium bond, attacks the partially positive carbon of the ketone. In the process, a pair of pi electrons of the carbonyl bond migrate to the carbonyl oxygen and negatively charged oxygen becomes associated with magnesium bromide, producing a magnesium alkoxide.

(2) The oxygen-magnesium complex is broken up upon acidic aqueous work-up as a lone pair of electrons belonging to the negatively charged oxygen of the alkoxide salt abstracts a proton from a hydronium ion-furnishing the alcohol product.

Ask the Chef:

(1) *Can an organolithium compound substitute for the Grignard reagent in the synthesis of triphenylmethanol?*

Chef: Yes. In fact, organolithium compounds are usually more reactive than Grignard reagents (due to the higher degree of ionic character of a carbon-lithium bond) in the synthesis of alcohols from aldehydes and ketones, leading to a greater product yield. In contrast to using a Grignard reagent, the use of an organolithium compound requires a different experimental setting, such as conducting the reaction under an inert atmosphere.

(2) *Can a Grignard reagent convert an aldehyde to its corresponding alcohol?*

Chef: Yes. In general, aldehydes are more reactive to the Grignard reagents nucleophilic attack than are ketones, for electronic and steric reasons. Bonded to only one alkyl group, the carbonyl carbon of the aldehyde is electron poor compared with ketones. Furthermore, containing only one bulky alkyl group, the carbonyl carbon of the aldehyde is more exposed to nucleophilic attack.

(3) *Will substituting anhydrous diethyl ether with another solvent, such as water, produce the same results?*

Chef: No. The Grignard reagent is highly basic, and acting as a strong Bronsted base, reacts readily with protic solvents such as water, alcohols, and amines.

Alkane
by Wolff-Kishner Reduction

Hexane

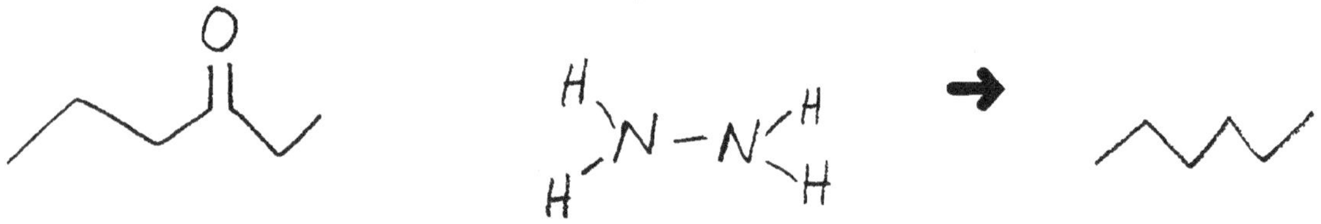

Propyl ethyl ketone   Hydrazine   Hexane

Ingredients:

Propyl ethyl ketone
Solution of hydrazine hydrate
Dilute hydrochloric acid
Potassium hydroxide (base)
Ethylene glycol (alcohol solvent)

Synthesis Synopsis:

Add Propyl ethyl ketone, hydrazine hydrate solution, and dilute
hydrochloric acid to a round-bottomed flask. Reflux. Add potassium
hydroxide and ethylene glycol. Distill to purify. Assess purity and yield
of product by gas chromatography.

Reaction Mechanism:
Wolff-Kishner Reduction

Step 1	
Step 2	
Step 3	
Step 4	
Resonance Structures:	
Step 5	
Step 6	

Reaction Mechanism:
Wolff-Kishner Reduction

Resonance Structures:	
Step 7	
Step 8	
Step 9	
Step 10	

Explanation of Steps:

(1) Nucleophilic hydrazine attacks the partially positive carbonyl carbon of the ketone, causing pi electrons to migrate to carbonyl oxygen.

(2) A reaction with dilute acid leads to fast proton transfer-- protonation at negatively charged oxygen and deprotonation at positively charged nitrogen.

(3) Protonation then occurs at the hydroxyl group (protonation of nitrogen leads back to the initial reactant) and a good leaving group is formed ($-OH_2^+$).

(4) A resonance stabilized intermediate is produced upon dissociation of $-OH_2^+$.

(5) A hydrazone is formed as positively charged nitrogen is deprotonated.

(6) Water is formed and a resonance stabilized intermediate is produced as hydroxide ion abstracts a proton from nitrogen.

(7) Water, acting as a Bronsted acid, yields a proton to negatively charged carbon (protonation of nitrogen would generate the starting compound), forming an azo intermediate.

(8) Nitrogen gains a negative charge as it is, once again, deprotonated by a hydroxide ion contributed by potassium hydroxide.

(9) Heterolytic cleavage occurs between carbon and nitrogen, displacing nitrogen gas and imparting a negative charge on carbon.

(10) The alkane product is formed upon protonation of the carbanion.

Ask the Chef:

(1) *Would the Clemmensen reduction method produce the same results?*

<u>Chef</u>: Yes. Clemmensen reduction is a suitable alternative method for the reduction of Propyl ethyl ketone. However, be advised that reducing the carbonyl group of an aldehyde or ketone to a methylene group by the Clemmensen reduction method is not always an alternative to Wolff-Kishner reduction. To prevent unwanted side reactions, compounds containing base-sensitive groups are not reduced by the Wolff-Kishner reduction method. Similarly, compounds containing acid-sensitive groups (e.g. an -OH group) are not reduced by the Clemmensen method of reduction (which uses an acidic solution of zinc dissolved in mercury as the reducing agent).

(2) *Is heat required?*

<u>Chef</u>: High temperature is necessary to facilitate deprotonation of the -NH$_2$ group of the hydrazone. However, Wolff-Kishner reduction can be carried out at room temperature when Potassium hydroxide and ethylene glycol are substituted by potassium *tert* butoxide and DMSO.

Aromatic Ketone
by Friedel-Crafts Acylation

1-Phenylethanone

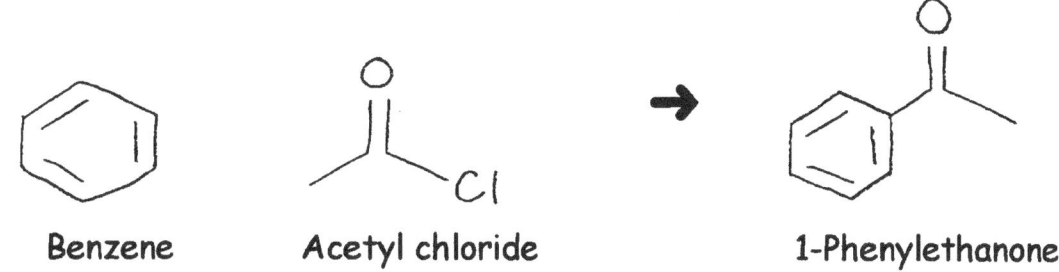

Benzene   Acetyl chloride   1-Phenylethanone

Ingredients:

Benzene (aromatic cycloalkene)
Acetyl chloride (acyl halide/Lewis base)
Aluminum Chloride (Lewis acid catalyst)
Water (for aqueous work-up)

Synthesis Synopsis:

Add benzene, acetyl chloride, and aluminum chloride to an Erlenmeyer flask. Heat flask on water bath. Cool solution. Add water to flask. Extract and dry the organic product. Record mass and melting point of precipitated product.

Reaction Mechanism:
Friedel-Crafts Acylation

Step 1	
Step 2	
Resonance Structures:	
Step 3	
Step 4	
Step 5	
Step 6	

Explanation of Steps:

(1) A lone pair of electrons from chlorine of acetyl chloride bonds with aluminum of aluminum chloride. In forming this complex, chlorine gains a positive charge.

(2) The complex is then dissociated, forming a tetrachloroaluminate anion ($-AlCl_4$) and an electrophilic, resonance-stabilized acylium ion.

(3) A cyclohexadienyl cation is produced as positively charged carbon of the acylium ion is attacked by pi electrons of the benzene ring.

(4) An acylbenzene is formed as a lone pair of electrons on chlorine of the tetrachloroaluminate anion abstracts a proton from the carbon bearing the acyl group.

(5) Unlike the acylium ion, the aluminum chloride (a weaker electrophile) does not react with the benzene ring. Rather, a lone pair of electrons of the newly formed acyl group bonds with the Lewis acid.

(6) Aqueous work-up serves to free the acylbenzene product from aluminum chloride by hydrolyzing the aluminum salts.

Ask the Chef:

(1) *Is multiple substitution a* problem?

<u>Chef</u>: No. The acyl substituent is electron withdrawing and renders the benzene ring inactive to further substitution.

(2) *Referring back to Step 1, why does aluminum chloride bind with chlorine of acetyl chloride rather than carbonyl oxygen?*

<u>Chef</u>: The isomeric complex between aluminum chloride and carbonyl oxygen exists in equilibrium with aluminum chloride bound to chlorine. However, it is the dissociation of the halogen of the acyl halide with aluminum chloride which leads to the electrophilic, resonance-stabilized acylium ion.

(3) *Referring to Step 2, what causes the dissociation of the acetyl chloride/aluminum chloride complex?*

<u>Chef</u>: Electron distribution to the more electronegative chlorine atom bearing a positive charge leads to a resonance stabilized intermediate.

Alkylbenzene
by Friedel-Crafts Alkylation

Ethylbenzene

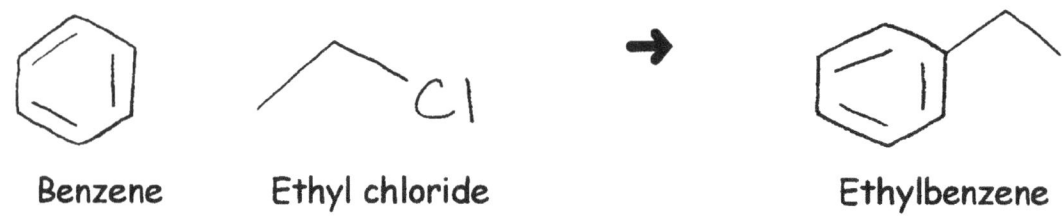

Benzene   Ethyl chloride   Ethylbenzene

Ingredients:

Benzene (aromatic cycloalkene)
Ethyl chloride (primary alkyl halide)
Aluminum chloride (Lewis acid catalyst)

Synthesis Synopsis:

Add benzene, ethyl chloride, and aluminum chloride to a round-bottomed flask. Reflux. Extract and dry the organic product. Distill to purify. Assess purity and yield of product by gas chromatography.

Reaction Mechanism:
Friedel-Crafts Alkylation

Step 1	
Step 2	
Step 3	

Explanation of Steps:

(1) The chlorine of ethyl chloride bonds with aluminum of aluminum chloride to form an electrophile suitable for attack by benzene.

(2) The partially positive, halogen-bearing carbon is attacked by benzene. In the process, the carbon-halogen bond is broken and a tetrachloroaluminate ion is formed.

(3) Hydrochloric acid and the alkylbenzene product are produced as a lone pair of electrons on chlorine of the tetrachloroaluminate ion abstracts a proton form the carbon bearing the alkyl substituent.

Ask the Chef:

(1) *Is polysubstitution a problem? If so, can it be* controlled?

<u>Chef</u>: Polysubstitution is a problem when introducing alkyl groups to a benzene ring by Friedel-Crafts alkylation. Friedel-Crafts alkylation introduces an electron-rich group which activates the ring. Activated rings are susceptible to further substitution. Polysubstitution can be controlled by using a blocking strategy (e.g., reverse sulfonation). The carbon bonded to a blocking group is protected from electrophilic substitution.

(2) *How do substituted benzene rings react to Friedel-Crafts alkylations* ?

<u>Chef</u>: It depends on the nature of the ring's substituent(s). Rings containing deactivating (electron withdrawing) groups are not susceptible to Friedel-Crafts alkylation. Furthermore, bulky alkyl substituents direct regioselectivity.

(3) *Referring to Step 2, does the ethyl benzene product compete with starting benzene for attack on the partially positive, halogen-bearing carbon?*

<u>Chef</u>: Yes. In fact, ethylbenzene is *more* reactive than benzene towards electrophilic attack. Therefore, excess benzene is used to increase the likelihood that the electrophilic species will react with benzene and not the alkylbenzene product.

β-Ketoester
by Claisen Condensation

Ethyl 3-oxobutanoate

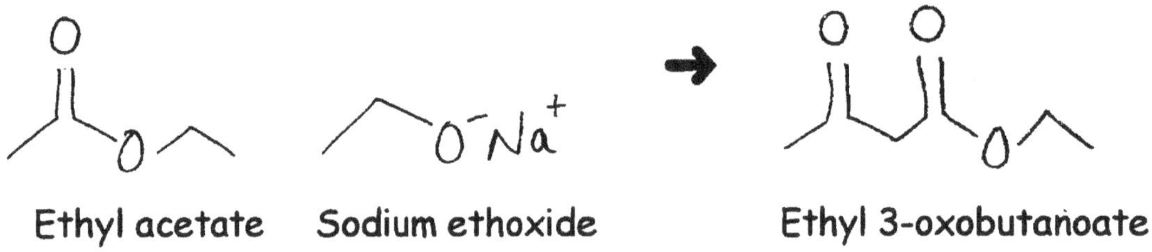

Ethyl acetate   Sodium ethoxide   Ethyl 3-oxobutanoate

Ingredients:

Ethyl acetate (ester)
Ethanol (alcohol solvent)
Sodium ethoxide (Bronsted base)
Dilute hydrochloric acid (for aqueous work-up)

Synthesis Synopsis:

Add ethyl acetate, sodium ethoxide, and ethanol to a round-bottomed flask. Reflux. Add dilute hydrochloric acid. Extract and dry the organic product and distill to purify. Assess purity and yield of product by gas chromatography.

Reaction Mechanism:
Claisen Condensation

Step 1	
Resonance Structures:	
Step 2	
Step 3	
Step 4	
Resonance Structures:	
Step 5	

Explanation of Steps:

(1) An alpha hydrogen of the ester is removed by ethoxide to furnish ethanol and a resonance stabilized enolate anion.

(2) The lone pair of electrons (previously bound to the alpha hydrogen) of carbon attacks the partially positive carbonyl carbon of ethyl acetate. In the process, pi electrons of the carbonyl bond migrate to oxygen.

(3) Negatively charged oxygen reforms the carbonyl bond. In the process, ethoxide is displaced and a β-ketoester is formed.

(4) Ethanol is regenerated and an enolate anion is produced as an alpha hydrogen of the β-ketoester is abstracted by ethoxide. Resonance forms of the enolate ion are produced as valence electrons are redistributed to offset the charge on carbon.

(5) Negatively charged carbon is protonated during aqueous work-up.

Ask the Chef:

(1) *Why must the starting ester contain two alpha hydrogens?*

Chef: The initial alpha hydrogen is required in the formation of the enolate anion of the starting ester. An additional alpha hydrogen is required in the formation of the ester enolate of the β-ketoester.

(2) *Referring to step 1, how favorable is the acid-base reaction between ethoxide and ethyl acetate?*

<u>Chef</u>: Ethanol, the conjugate acid of ethoxide, has a lower pK_a than ethyl acetate, i.e., ethanol is a stronger acid than ethyl acetate Therefore, initial deprotonation of the ester by ethoxide is rather unfavorable. The equilibrium of the reaction is shifted upon β-ketoester formation.

(3) *Referring to Step 4, how favorable is the acid-base reaction between ethoxide and the β-ketoester?*

<u>Chef</u>: Flanked by *two* carbonyl groups, the alpha hydrogens of the β-ketoester are more acidic than the ester. Consequently, unlike the preceding steps, Step 4 is virtually irreversible due to the favorable acid-base reaction between ethoxide and the β-ketoester (which has a lower pK_a value than the conjugate acid of ethoxide).

Amine
by Hofmann Rearrangement

Aniline

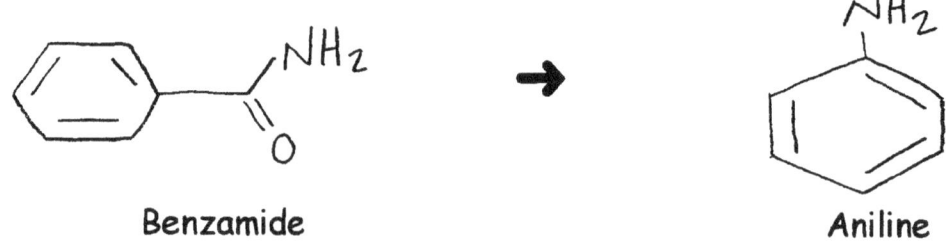

Benzamide → Aniline

Ingredients:

Bromine
Benzamide (primary amide)
Concentrated sodium hydroxide (base)

Synthesis Synopsis:

Add bromine, benzamide, and sodium hydroxide to an Erlenmeyer flask. Heat flask gently over hot plate, stirring occasionally. Set aside mixture and allow to cool. Assess purity and yield of product by gas chromatography.

Reaction Mechanism:
Hofmann Rearrangement

Step 1	$\rightarrow$ H_2O +
Resonance Structures:	$\longleftrightarrow$
Step 2	$\rightarrow$ Br–Br $\rightarrow$ $+Br^-$
Step 3	$\rightarrow$ H_2O +
Resonance Structures:	$\longleftrightarrow$
Step 4	$\rightarrow$ $N=C=O$ $+Br^-$
Step 5	$\rightarrow$

Reaction Mechanism:
Hofmann Rearrangement

Resonance Structures:	
Step 6	
Step 7	
Step 8	
Step 9	

Explanation of Steps:

(1) A resonance stabilized amidate ion (a strong nucleophile) is produced as a hydroxide ion abstracts a proton from nitrogen of the primary amide.

(2) Negatively charged nitrogen is halogenated as the amidate ion attacks electrophilic bromine (a good leaving group), to form the N-bromo amide.

(3) A resonance stabilized N-bromo amide anion is formed as a second proton is abstracted by hydroxide ion.

(4) Upon **rearrangement**, an isocyanate intermediate (a species in which all atoms have complete valence shells) is produced: bromine (a good leaving group) is cleaved heterolytically (departing with the electrons that made up the bond between it and nitrogen) as the alkyl group migrates (with its bonding electrons) to nitrogen.

(5) The highly electrophilic carbonyl carbon is attacked by hydroxide ion, producing a resonance stabilized intermediate.

(6) A carbamic acid (an unstable species easily decarboxylated by base) is produced upon proton transfer between water (acting as a Bronsted acid) and negatively charged nitrogen.

(7) Deprotonation occurs at carbamic acid's hydroxyl group.

(8) Carbon dioxide is released as negatively charged oxygen forms a pi bond with adjacent carbon.

(9) The amine is produced as water yields a proton to negatively charged nitrogen.

Ask the Chef:

(1) *Referring to Step 4, what prompts rearrangement?*

<u>Chef</u>: Deprotonation of the N-bromo amide by hydroxide ion sets up an anionic compound (N-bromo amide anion) with characteristics that favor rearrangement: negatively charged nitrogen is bonded to an excellent leaving group (bromine) and has a lone pair of electrons that can form a pi bond to the adjacent carbon that bears a potential migrating alkyl group.

(2) *How favorable is the reaction between the isocyanate intermediate and the hydroxide ion?*

<u>Chef</u>: Flanked by two highly electronegative heteroatoms, the doubly bonded carbon of the isocyanate intermediate is rendered considerably electrophilic and reacts readily with the nucleophilic hydroxide ion.

Alkene
by Hofmann Elimination

1-Butene

2-Butanamine   1-Butene

Ingredients:

2-Butanamine (amine)
Iodomethane (alkyl halide)
Silver oxide in water solution (source of -OH⁻)

Synthesis Synopsis:

Add 2-butanamine and iodomethane to an Erlenmeyer flask. Heat flask gently on hot plate, stirring occasionally. Remove flask from heat and add silver oxide solution. Heat and stir. Distill to purify. Assess purity and yield of product by gas chromatography.

Reaction Mechanism
Hofmann Elimination

Step 1	
Step 2	
Step 3	

Explanation of Steps:

(1) In an S_N2 reaction, 2-butanamine attacks iodomethane. In the process, nitrogen gains a positive charge as it is methylated and a nucleophilic iodine ion is produced.

(2) Hydrogen iodide is produced as the iodine ion deprotonates positively charged nitrogen.

(3) Steps 1 and 2 are repeated until nitrogen is completely methylated in a process known as *exhaustive methylation*. Consequently, a bad leaving group ($-NH_2$) is converted into a moderately good leaving group ($-N(CH_3)_3^+$). In an E2 reaction (simultaneous proton transfer to base, double bond formation, and ejection of leaving group) facilitated by heating, alkene product is formed.

Ask the Chef:

(1) *Why does the hydroxide ion deprotonate the β-carbon bonded to the most hydrogens?*

<u>Chef</u>: Contrary to Saytzev elimination, the least highly substituted product commonly predominates the Hofmann elimination. In Hofmann elimination, the more stable transition state involves the less substituted carbanion, whereas in Saytzev elimination, the more stable transition state leading to the alkene product places the negative charge on the more highly substituted carbon. This pathway is typical for E2 reactions involving poor leaving groups. In Hofmann elimination, the leaving group is a tertiary amine, which does not dissociate easily (hence the requirement of heat). Consequently, when a base begins to abstract a proton from the β-carbon (the carbon adjacent to the carbon bearing the leaving group), a negative charge develops. Contrary to a carbocation transition state, the carbanion is stabilized by the number of electron *withdrawing* groups attached to the carbon bearing the negative charge (i.e., the less electron donating alkyl groups attached to the proton yielding carbon (the β-carbon), the more stable the carbanion). Therefore, the route to the major product involves deprotonation at the less substituted carbon.

α-Haloacid
by the Hell-Volhard-Zelinsky Reaction

2-Bromobutanoic acid

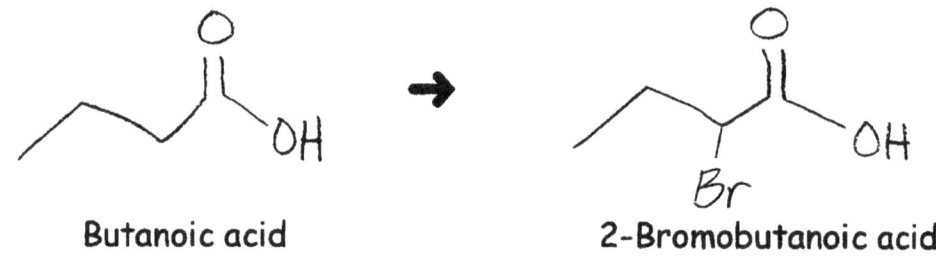

Butanoic acid   2-Bromobutanoic acid

Ingredients:

Bromine
Red Phosphorus
Butanoic acid (carboxylic acid)
Water (for aqueous work-up)

Synthesis Synopsis:

Add bromine and red phosphorus to an Erlenmeyer flask. Heat flask gently over hot plate. Add butanoic acid and continue heating. Add water. Distill to purify.

Reaction Mechanism:
Hell-Volhard-Zelinsky Reaction

Step 1	
Step 2	
Step 3	
Resonance Structures:	
Step 4	
Step 5	
Step 6	

Reaction Mechanism:
Hell-Volhard-Zelinsky Reaction

Step 7	
Step 8	
Step 9	

Explanation of Steps:

(1) A bromine ion is displaced and a moderately good leaving group is formed ($-HOPBr_2^+$), as the carboxylic acid attacks PBr_3 in an S_N2 fashion.

(2) An acyl bromide (in keto form) is produced as nucleophilic bromine ion displaces $-HOPBr_2^+$ in an S_N2 fashion.

(3) A resonance stabilized intermediate is produced upon protonation of carbonyl oxygen.

(4) The enol form of the acyl halide is produced as an enolizable hydrogen dissociates heterolytically and a double bond forms in a reversible process known as *keto-enol-tautomerism*.

(5) The alpha carbon is brominated and a carbonyl bond forms as the nucleophilic enol attacks bromine.

(6) An α-bromo acyl bromide is produced as positively charged oxygen is deprotonated by bromide ion.

(7) Upon aqueous work-up, water adds to carbonyl carbon.

(8) Bromine acts as the leaving group as the carbonyl bond is reformed.

(9) The α-haloacid is produced as the bromide ion deprotonates oxygen.

Ask the Chef:

(1) *Referring to Step 3, what species serves to yield hydrogen ions?*

<u>Chef</u>: As can be seen, for every time PBr_3 yields Br (Step 1) it gains an OH in return (Step 2). Therefore, upon yielding its three bromine atoms, PBr_3 becomes hydrogen yielding H_3PO_3 (phosphorus acid).

(2) *How is PBr_3 produced?*

<u>Chef</u>: Phosphorus tribromide is a product of the reaction between red phosphorus and bromine. It is produced in situ due to its highly corrosive nature.

Alcohol
by Oxymercuration-Demercuration

1-Methylcyclopentanol

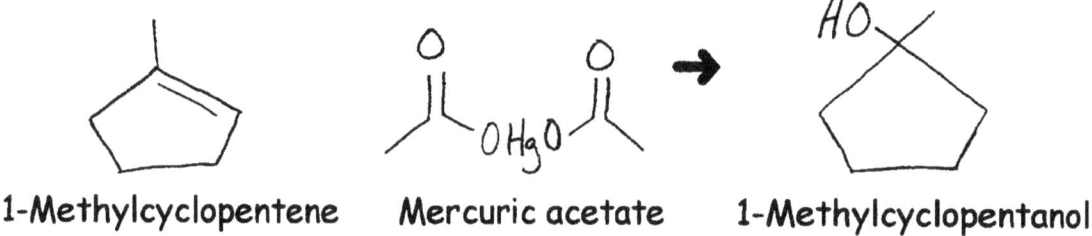

1-Methylcyclopentene   Mercuric acetate   1-Methylcyclopentanol

Ingredients:

Water (nucleophilic solvent)
Mercuric acetate in THF
1-Methylcyclopentene (alkene)
Sodium borohydride (reducing agent)

Synthesis Synopsis:

Add water, 1-Methylcyclopentene, and mercuric acetate in THF solution to a round-bottomed flask. Reflux. Add sodium borohydride. Record mass and melting point of precipitated product.

Reaction Mechanism:
Oxymercuration-Demercuration

Step 1	
Step 2	
Step 3	
Step 4	
Step 5	

Explanation of Steps:

(1) Mercuric acetate dissociates heterolytically into $^+HgOCOCH_3$ (a Lewis acid) and CH_3OCO^- (an acetate anion and Lewis base).

(2) A bridged mercurium ion intermediate is formed as pi electrons attack positively charged mercury of the Lewis acid.

(3) Water, acting as a nucleophile, opens the ring by attacking the more substituted carbon of the three-membered bond.

(4) An organomercurial alcohol is formed as positively charged oxygen is deprotonated by water.

(5) Demercuration occurs as a hydride ion (donated by sodium borohydride) replaces the acetoxymercuri group of the organomercury alcohol.

Ask the Chef:

(1) *Wouldn't acid catalyzed hydration of the alkene produce the same results?*

Chef: Not necessarily. Carbocations formed during the acid-catalyzed hydration of alkenes lead to rearrangements. There is no carbocation formation in oxymercuration-demercuration and so rearrangement does not occur.

(2) *Is the positive charge evenly distributed over all three atoms of the bridged mercurium ion intermediate? If so, why is the more highly substituted carbon the subject of nucleophilic attack?*

<u>Chef</u>: The atoms of the three membered ring share a fraction of the positive charge. Water's attack at the more highly substituted carbon leads to a more stable transition state, placing the partial positive charge on a secondary rather than a primary carbon.

(3) *Would the use of an alcohol solvent in place of water produce the same results?*

<u>Chef</u>: No. Substituting for water, an alcohol solvent would serve as the nucleophilic species and attack the mercurium ion (as water does in Step 3 of the reaction) to produce the corresponding *ether*.

Alcohol
by Hydroboration-Oxidation

1-Hexanol

1-Hexene   1-Hexanol

Ingredients:

1-Hexene (alkene)
Borane in THF solution (electrophile)
Sodium hydroxide (base)
Hydrogen peroxide (oxidizing agent)

Synthesis Synopsis:

Add 1-hexene and borane in THF solution to a round-bottomed flask.
Relfux. Add hydrogen peroxide and continue reflux. Add sodium
hydroxide. Distill to purify.

Reaction Mechanism:
Hydroboration-Oxidation

Step 1	
Step 2	
Step 3	
Step 4	

Explanation of Steps:

(1) In a concerted reaction, borane (a strong electrophile) is attacked by the pi electrons of the alkene. Through a four-center transition stage, hydrogen and BH_2 are added to the double bond in an anti-Markovnikov manner (hydrogen is added to the more substituted carbon). The alkylborane formed reacts with two more molecules of alkene to form a trialkylborane.

(2) A lone pair electrons on the hydroperoxide ion attacks boron of the trialkylborane.

(3) An alkyl group migrates from the negatively charged boron to oxygen, ejecting hydroxide ion in the process. This step is repeated two more times, forming a trialkylborate.

(4) The trialkylborate reacts with sodium hydroxide, furnishing the alcohol product and sodium borate.

Ask the Chef:

(1) *Wouldn't acid catalyzed hydration of the starting alkene produce the same results?*

Chef: Unlike acid catalyzed hydration, hydroboration-oxidation adds water to a double bond in an anti-Markovnikov manner.

(2) *Why is anti-Markovnikov hydration favored, i.e., why isn't hydrogen added to the less highly substituted end of the double bond?*

Chef: The addition of borane in an anti-Markovnikov orientation furnishes a more stable transition state in which the more highly substituted carbon of the double bond bears a partial positive charge. The electron withdrawing power of electrophilic boron is felt across the double bond as pi electron density concentrates at the less highly substituted carbon. The transition state involving an electron deficiency at the more highly substituted carbon is stabilized by the adjacent electron rich alkyl group.

Amine
by Reductive Amination

N,N-Dimethycyclohexylamine

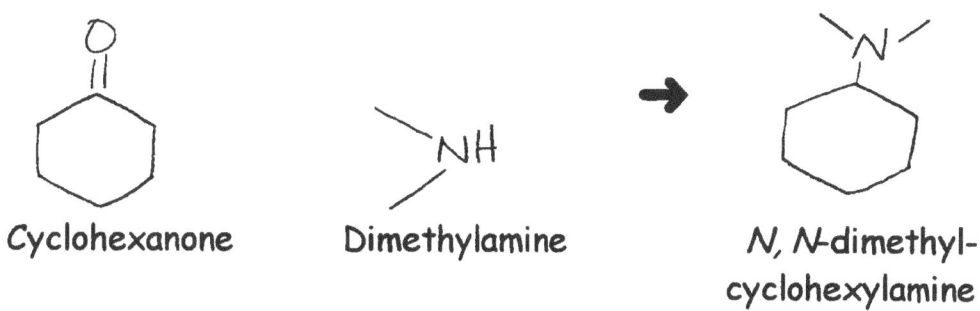

Cyclohexanone   Dimethylamine   N, N-dimethyl-
cyclohexylamine

Ingredients:

Dimethylamine (secondary amine)
Cyclohexanone (ketone)
Ethanol (solvent)
Sodium cyanoborohydride (reducing agent)

Synthesis Synopsis:

Add dimethylamine, cyclohexanone, and ethanol to a round-bottomed
flask. Reflux. Add sodium cyanoborohydride. Extract and dry the
organic product and distill to purify. Assess purity and yield of product
by gas chromatography.

Reaction Mechanism:
Reductive Amination

Step 1	
Step 2	
Step 3	
Step 4	
Step 5	
Step 6	

Explanation of Steps:

(1) A lone pair of electrons on dimethylamine attacks the carbonyl carbon as pi electrons migrate to oxygen. In the process nitrogen gains a positive charge.

(2) A hemiaminal is produced as proton transfer occurs between negatively charged oxygen and positively charged nitrogen.

(3) The hydroxyl group is protonated, turning a bad leaving group (-OH) into a good leaving group ($-OH_2^+$).

(4) The lone pair of electrons on nitrogen forms a double bond with carbon as $-OH_2^+$ leaves. In the process, nitrogen gains a positive charge.

(5) An enamine (an α-β unsaturated tertiary amine) is formed as an alpha carbon loses a proton and pi electrons migrate to the positively charged nitrogen.

(6) Upon hydride transfer, sodium cyanoborohydride reduces the enamine to its corresponding amine.

Ask the Chef:

(1) *Can sodium borohydride be used in place of sodium cyano-borohydride?*

Chef: Not if one chooses to reduce the enamine as it is formed. Sodium cyanoborohydride is employed because it is a less reactive reducing agent than sodium borohydride. Unlike sodium cyanoborohydride, sodium borohydride will react with the starting ketone.

(2) *Referring to Step 3, which species serves to protonate oxygen?*

Chef: The alcohol solvent provides the hydrogen ions. Upon protonation, ethanol is converted to ethoxide ion. Ethanol is regenerated upon transfer of an alpha hydrogen to ethoxide ion.

(3) *Would the use of ammonia in place of dimethylamine produce the same results?*

Chef: No. Containing two alkyl groups, dimethylamine leads to a tertiary amine. Similarly, the reaction between cyclohexanone and ammonia (an amine containing no alkyl groups) leads to cyclohexanamine, a primary amine.

Ether
by Williamson Ether Synthesis

Ethyl propyl ether

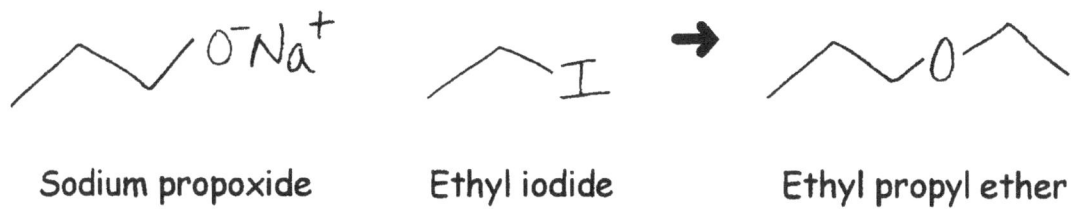

Sodium propoxide   Ethyl iodide   Ethyl propyl ether

Ingredients:

Ethyl iodide (alkyl halide)
Sodium propoxide (alkoxide salt)
Dimethyl sulfoxide (DMSO) (polar aprotic solvent)

Synthesis Synopsis:

Add sodium propoxide, DMSO, and ethyl iodide to a round-bottomed
flask. Reflux. Extract and dry organic product. Distill to purify
Assess purity and yield of product by gas chromatography.

Reaction Mechanism:
Williamson Ether Synthesis

Step 1	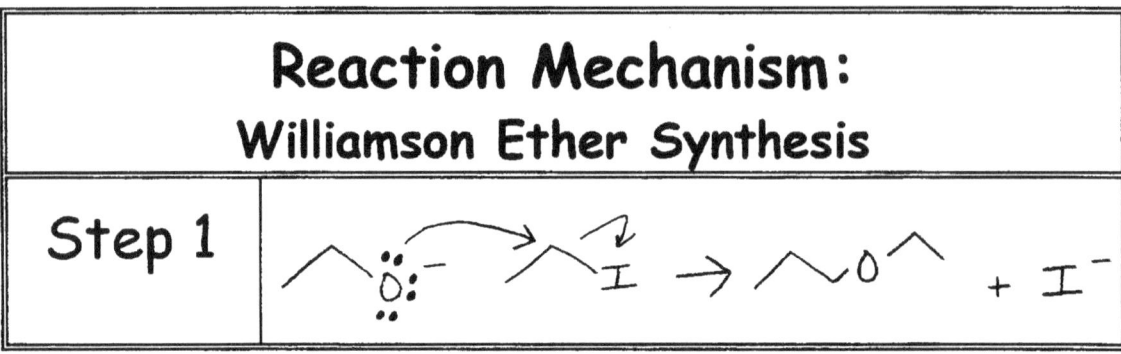

Explanation of Steps:

(1) In a typical S_N2 fashion, the nucleophilic alkoxide ion displaces the iodide ion from the alkyl halide, forming the ether product.

Ask the Chef:

(1) *The above reaction involves ethyl iodide, a primary alkyl halide. Would the use of a secondary or tertiary alkyl halide produce the same results?*

Chef: Steric hindrance, a factor in the rate of an S_N2 reaction, would slow a reaction involving a secondary alkyl halide substrate and lead to a negligible amount of product when a tertiary alkyl halide is employed.

(2) *Would substituting DMSO with a polar protic solvent, such as water, produce the same results?*

Chef: No. Ion-dipole interaction between water molecules and the negatively charged alkoxide nucleophile would impede the S_N2 process. Similarly, the reaction would not take place in a nonpolar solvent in which the starting alkoxide salt would not dissolve.

Alkyl Halide
by Allylic Bromination

3-Bromocyclohexene

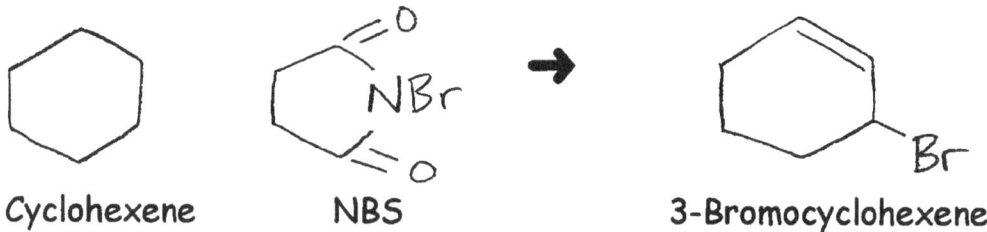

Cyclohexene   NBS   3-Bromocyclohexene

Ingredients:

Cyclohexene (alkene)
N-Bromosuccinimide (NBS)
Carbon tetrachloride (solvent)
Hydrogen bromide[*]

Synthesis Synopsis:

Add cyclohexene, NBS, hydrogen bromide, and carbon tetrachloride to an Erlenmeyer flask. Heat gently over hot plate. Distill to purify. Assess purity and yield of product by gas chromatography.

[*]A trace amount of hydrogen bromide is employed for this reaction.

Reaction Mechanism:
Allylic Bromination

Step 1	
Step 2	
Step 3	
Resonance Structures:	
Step 4	

Explanation of Steps:

(1) The nitrogen of NBS reacts with hydrogen bromide to yield succinimide and bromine.

(2) Bromine dissociates into bromine radicals.

(3) A bromine radical abstracts an allylic hydrogen, producing a resonance stabilized allylic radical.

(4) The alkyl halide product is formed and a bromine radical is regenerated as bromine is cleaved homolytically and attacked by the allylic radical.

Ask the Chef:

(1) *Referring to Step 4, why doesn't the allylic radical combine with a neighboring allylic radical?*

<u>Chef</u>: Such a reaction is observed but the product yield of such is negligible due to the more energetically favorable bond formation between the allylic and bromine radical.

(2) *What role does hydrogen bromide play?*

<u>Chef</u>: A trace amount of hydrogen bromide is used to react with NBS, which serves as an indirect source of bromine radicals. The reaction between NBS and hydrogen bromide produces Br_2. Using a trace amount of hydrogen bromide maintains a steady, but low, concentration of Br_2.

(3) *Why is NBS used in place of bromine as a source of bromine radicals?*

<u>Chef</u>: Generating bromine through the reaction of NBS and hydrogen bromide maintains a low and steady concentration of bromine, deterring addition to the double bond and the formation of the corresponding vicinal dibromide--which forms when a high concentration of bromine is present.

(4) *Is heat required?*

<u>Chef</u>: Yes. Heat serves to dissociate the bromine into bromine radicals. The role of heat may be played by light or a radical initiator (such as benzoyl peroxide) as well.

(5) *When using branched alkenes as the starting material, how regioselective is the bromine radical to the allylic position?*

<u>Chef</u>: Steric factors contribute to the selectivity of the bromine radical, increasing the likelihood of addition to the less substituted allylic carbon.

Ester
by Baeyer-Villiger Oxidation

Ethyl acetate

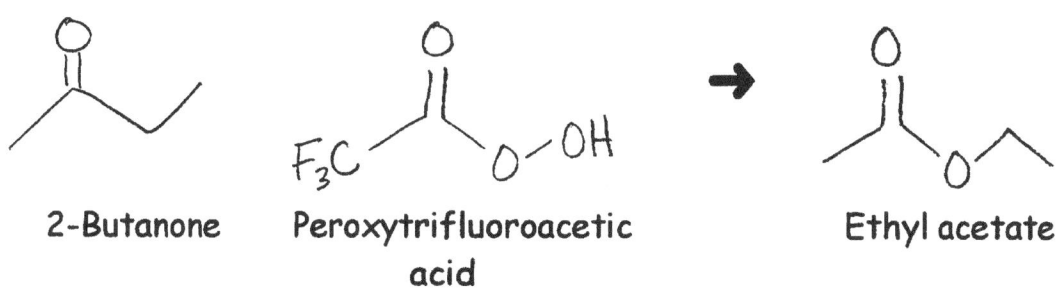

2-Butanone   Peroxytrifluoroacetic   Ethyl acetate
acid

Ingredients:

2-Butanone (ketone)
Peroxytrifluoroacetic acid (oxidizing agent)
Dichloromethane (solvent)

Synthesis Synopsis:

Add 2-Butanone, peroxytrifluoroacetic acid, and dichloromethane to an Erlenmeyer flask. Heat flask gently over hot plate. Remove flask from heat and set aside to cool. Assess purity and yield of product by gas chromatography.

Reaction Mechanism:
Baeyer-Villiger Oxidation

Step 1	
Step 2	
Step 3	

Explanation of Steps:

(1) Proton transfer occurs between the peroxy acid and the carbonyl group of the ketone. Such protonation actuates the carbonyl group toward nucleophilic attack by the negatively charged terminal end of the peroxy acid.

(2) Pi electrons of the carbonyl bond migrate to positively charged oxygen as the peroxy acid anion adds to the electrophilic carbonyl carbon.

(3) In a cyclic transition state, intramolecular proton transfer occurs between hydroxyl oxygen and oxygen of the carbonyl group. The carboxylic acid portion which results from the proton transfer is a good leaving group. The ester is formed through a concerted process involving heterolytic cleavage of the weak oxygen-oxygen bond and migration of the ethyl group to the increasingly electron deficient oxygen being created through dissociation.

Ask the Chef:

(1) *Referring to Step 3, couldn't the neighboring methyl group serve as the electron donor and migrate to the electron deficient oxygen in place of the primary alkyl (ethyl) group?*

Chef: Baeyer-Villiger oxidation involves the insertion of an oxygen atom between the carbonyl carbon and an alpha carbon atom (a carbon directly bonded to carbonyl carbon) in the formation of an ester. Positioning of the oxygen depends upon the tendency of the alpha carbon to migrate (known as migratory aptitude). Experiments have shown that the more highly substituted carbon migrates preferentially. Accordingly, primary alkyl groups have a greater tendency to migrate than methyl groups.

(2) *Is Baeyer-Villiger oxidation a useful method for producing esters from aldehydes?*

Chef: Baeyer-Villiger oxidation of aldehydes involves the migration of hydrogen rather than a neighboring alkyl group, in turn, producing the corresponding carboxylic acid.

www.ingramcontent.com/pod-product-compliance
Lightning Source LLC
Chambersburg PA
CBHW081143170526
45165CB00008B/2777